WHALES

What's inside?

TWO CAN

World of whales

Whales are enormous animals that swim, dive and play in the sea. They cannot breathe underwater, so they come up to the surface to take in fresh air. A whale has rubbery flippers and a powerful tail to help it swim.

Guess what?
Under a whale's skin, there is a layer of fat, called blubber. This keeps the whale warm in cold water.

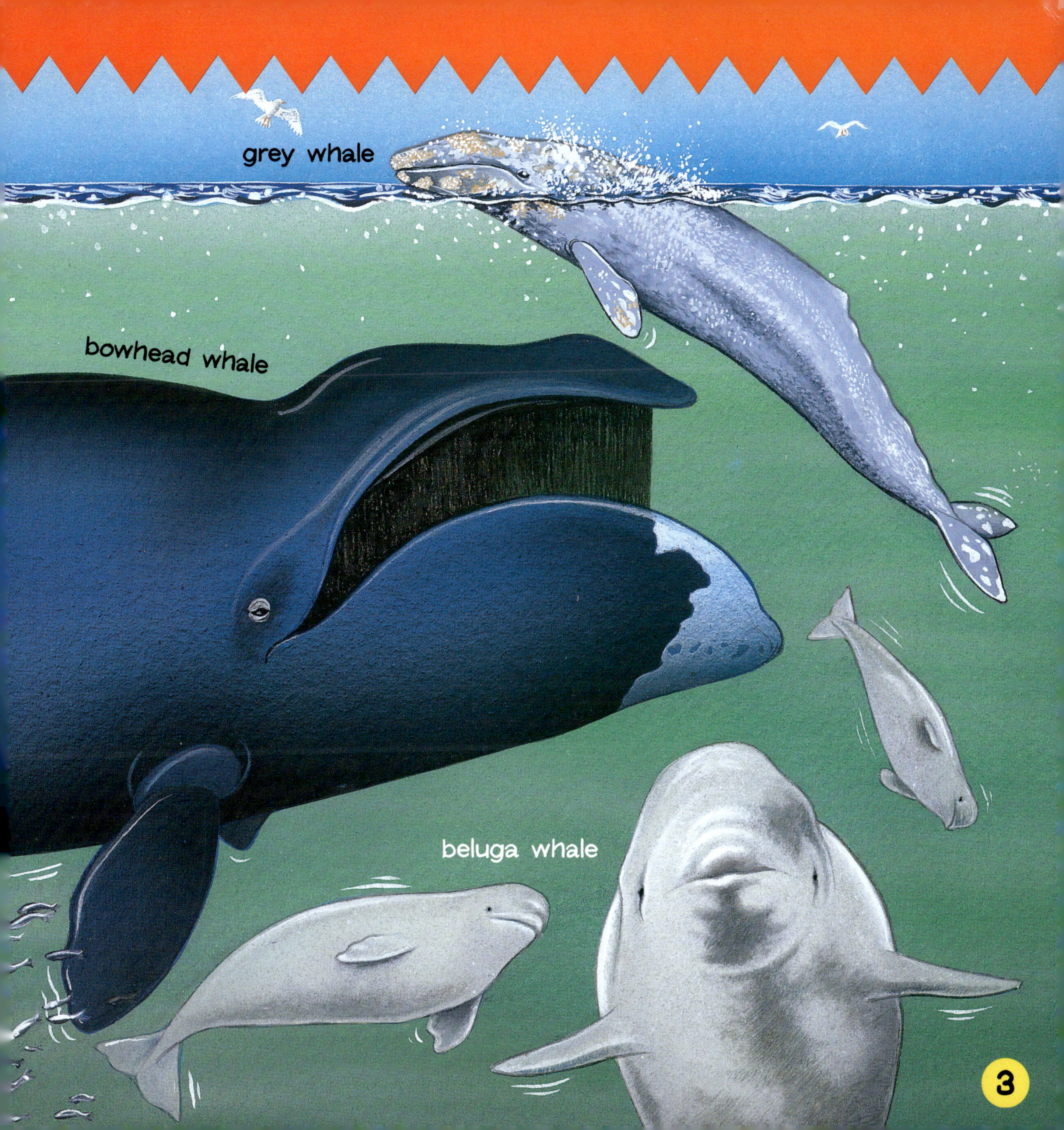
grey whale
bowhead whale
beluga whale

Blue whale

The blue whale is the largest animal in the world. It lives mainly in icy seas. As this gentle giant swims along, it gulps down mouthfuls of food. The blue whale loves to feast on krill, which are tiny creatures that float in the water.

How long?

A few blue whales have grown to 30 metres long. That's the same length as an aeroplane.

ha ha
What do whales chew?
Blubber gum!

Blue whale facts

A blue whale's tongue is so heavy that it would take 35 strong people to lift it up.

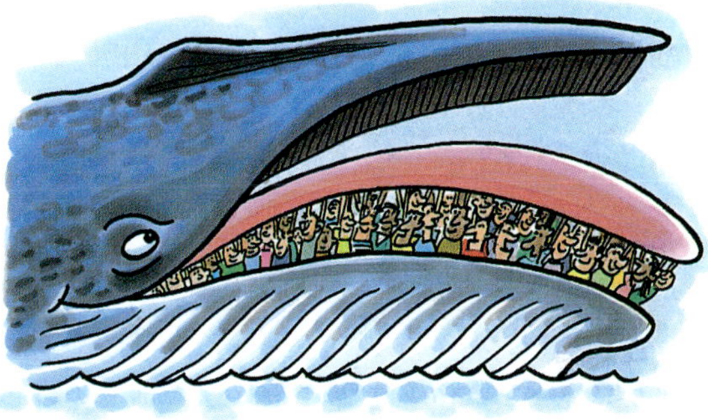

A blue whale can race along as quickly as a motorboat, but usually it prefers to swim much more slowly.

Small creatures, called barnacles, glue themselves on to a blue whale's skin. They ride on the whale as it looks for food, then they join in the feast!

Big splash

A whale spends most of its time underwater, but sometimes it swims to the top of the ocean to splash around in the waves. Then it takes a deep breath and dives back down.

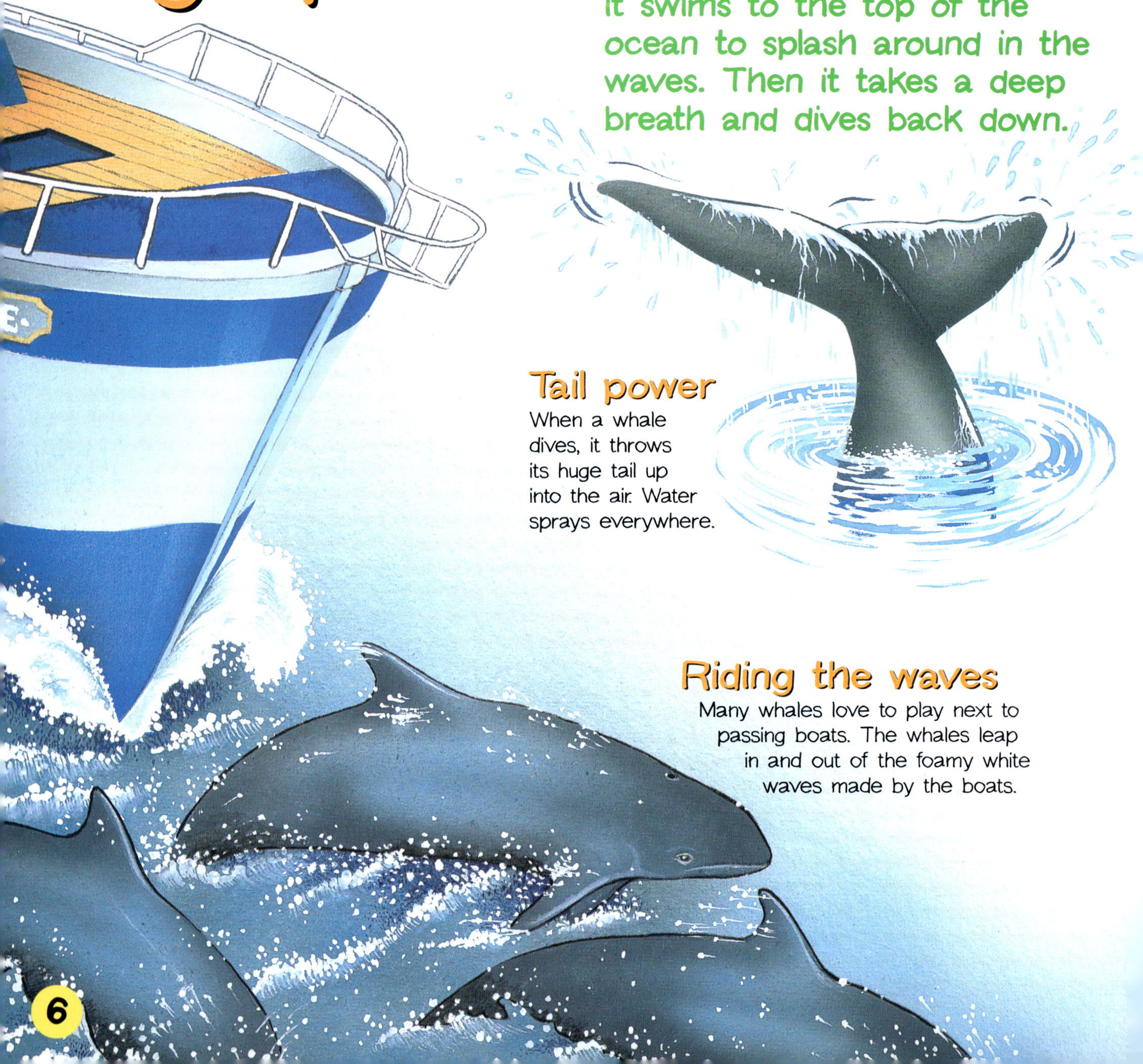

Tail power

When a whale dives, it throws its huge tail up into the air. Water sprays everywhere.

Riding the waves

Many whales love to play next to passing boats. The whales leap in and out of the foamy white waves made by the boats.

Take a deep breath

A whale breathes in air through one or two blowholes on the top of its head. When the whale breathes out, it shoots a misty spray high into the sky. This spray is called a spout.

Guess what?
Sperm whales can hold their breath underwater for up to two hours.

There she blows!

Many whales blow tall, thin spouts, while others blow short, fat ones. Some whales even blow two spouts!

Humpback whale

A humpback whale is a champion acrobat. Its favourite trick is to jump high into the air, then land with a tremendous splash. A humpback whale's large, knobbly head and long flippers make it easy to spot.

ha ha

How would a whale travel on land? By whale-way!

How heavy?

Keep out of the way of a mighty humpback whale. It's as heavy as three fire engines!

Humpback whale facts

A humpback whale likes to laze around and watch the world go by. It floats on its back, with its tail and flippers in the air.

Nosy humpbacks may swim up to a boat to see what's going on. The whales are so big, they often give the people on board a fright.

A male humpback whale is a fantastic crooner. He sings songs to his mate that can be heard for miles around!

Feed me!

Many whales have sharp teeth for catching and crunching up large sea animals. Other whales feed on tiny sea creatures, called krill. Instead of teeth, these whales have hundreds of bony strips in their mouths, called baleen.

Guess what?
Baleen is made of exactly the same stuff as your fingernails.

A big appetite

A minke whale takes a huge mouthful of water and krill. When it shuts its mouth, the water pours back out through the baleen, leaving the tasty food trapped inside.

Surfing for a snack

A fierce killer whale surfs towards the shore in search of dinner. The hungry whale hopes to snatch a plump sea lion with its sharp teeth.

A fishy trick

A humpback whale uses its blowholes to trap fish in a bubbly net. Then the whale swims towards the fish and gobbles them up.

Sperm whale

The sperm whale is a huge and daring hunter. It dives deep into the ocean to catch lobsters, giant squid and even fierce sharks. A sperm whale grabs its prey with its long, pointed teeth and gulps the animal down.

ha ha

What do baby whales play with? Doll-phins!

How deep?

A sperm whale can dive as deep as a submarine. The whale searches for food in the dark, murky water near the bottom of the ocean.

Sperm whale facts

A sperm whale's favourite food is giant squid, but it isn't a fussy eater. This whale will munch coconuts, apples and even stones or boots!

Sperm whales have contests to prove who is stronger. They bash their heads together until one of the whales gives in.

Every year, a father sperm whale leaves his family and swims thousands of miles to look for food. Sometimes, the father whale never comes back.

Baby whales

A mother whale gives birth to a baby, called a calf. She feeds it and keeps it safely by her side for about a year, until the calf is strong enough to look after itself.

Baby's first breath

As soon as a calf is born, its mother nudges it gently to the surface. Here, the baby whale takes its first breath.

Guess what?

A blue whale calf weighs almost two tonnes. That's as heavy as 550 newborn human babies!

Babysitting whales

Adult pilot whales form a ring around their calves to protect them from danger on all sides.

Play fight

Young male narwhals cross their long, spiralled tusks like swords and have play fights. This is how they learn to defend themselves.

Marvellous milk

A mother whale feeds her calf with milk from her body. This helps the calf to grow quickly. A baby fin whale can drink more than 600 litres of milk in just one day!

Best friends

Many whales live in groups, called herds. They travel together on long journeys and even talk to one another in their own special language.

Keep in touch

Whales talk to each other by making clicks and whistles. These sounds can travel through the water for up to 80 kilometres.

To the rescue

When a whale is sick or hurt, the other whales in the herd come to the rescue. They help the sick whale swim to the surface for air.

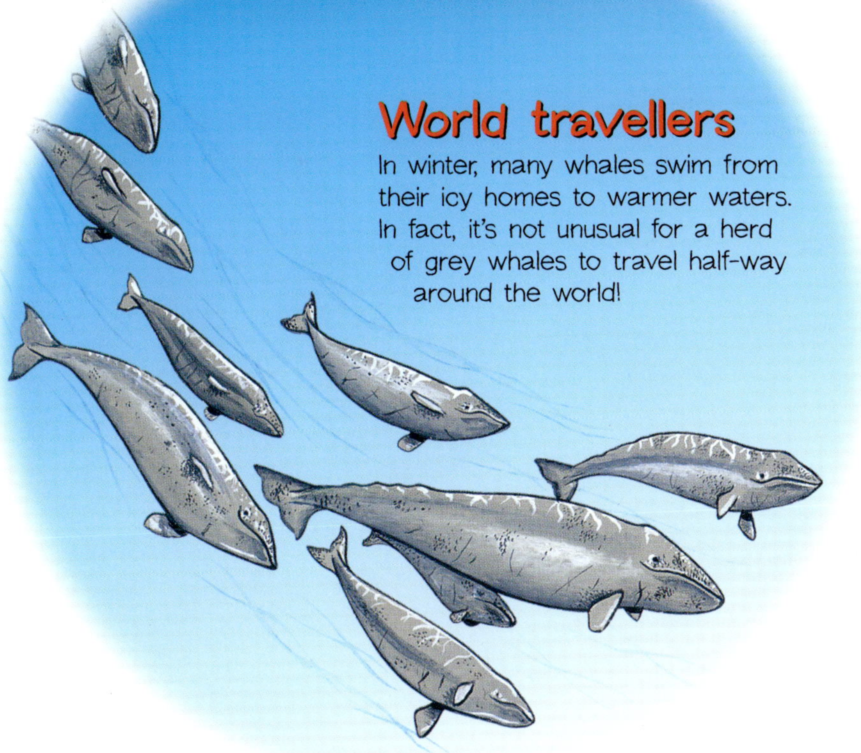

World travellers

In winter, many whales swim from their icy homes to warmer waters. In fact, it's not unusual for a herd of grey whales to travel half-way around the world!

Guess what?
Whales say hello by rubbing their heads together and patting each other with their flippers.

Lost and stranded

When a lost whale swims too close to shore, it may get stuck. Usually the unhappy whale has to wait for people to help free it before it can swim back to its friends.

Thousands of years ago, there were lots of whales swimming in the sea. Over time they began to die out. What went wrong?
SAVING THE WHALES
Where is everybody?
1
In the 1800s, people hunted and caught about 10,000 whales a year. These were mostly slow-moving sperm whales, hunted for their oil.
2
oil lamps
People sold different parts of the whales to make all kinds of things, from candles to lipstick.
face cream
candles
lipstick

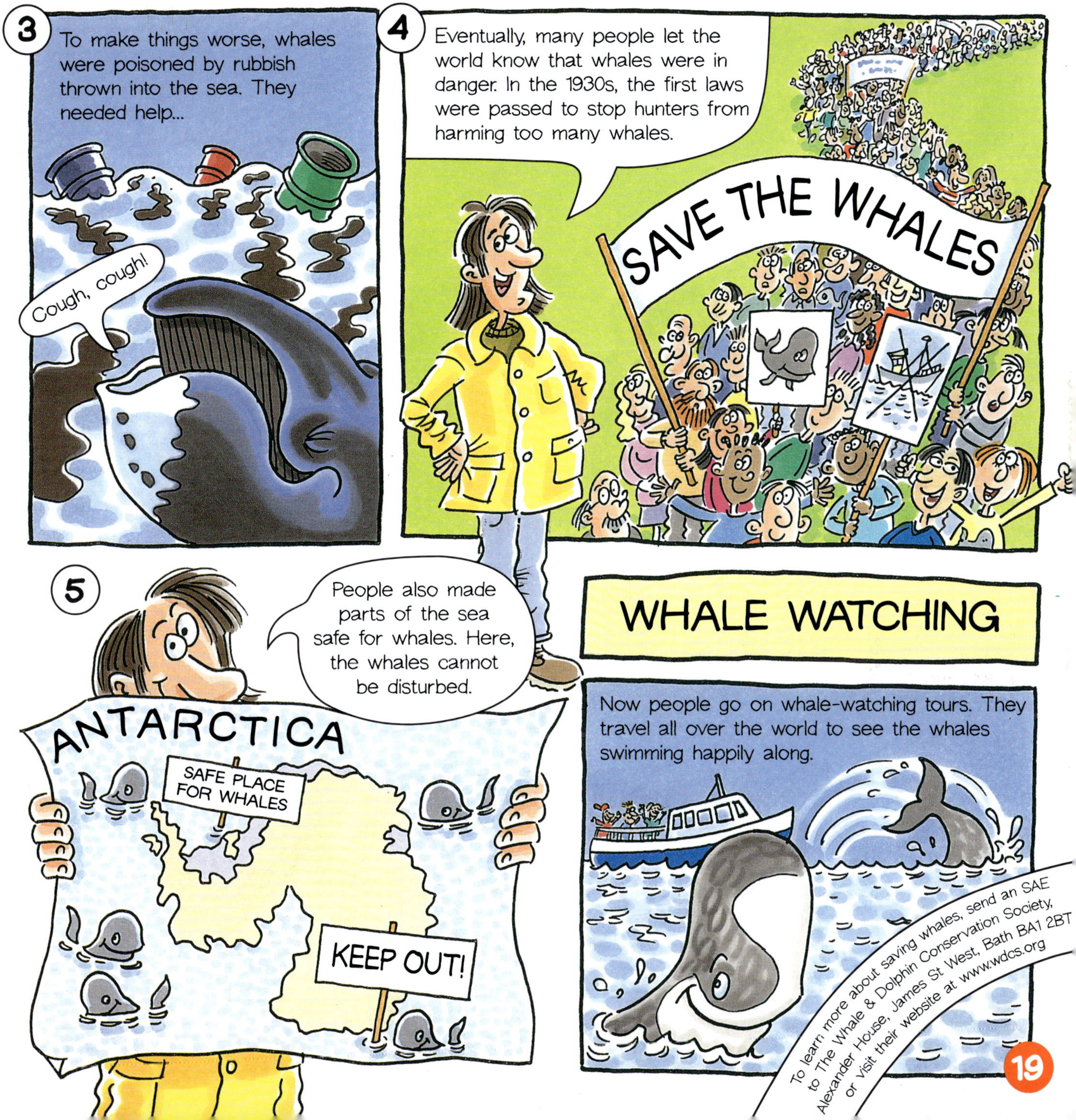
3
To make things worse, whales were poisoned by rubbish thrown into the sea. They needed help...
Cough, cough!
4
Eventually, many people let the world know that whales were in danger. In the 1930s, the first laws were passed to stop hunters from harming too many whales.
SAVE THE WHALES
5
People also made parts of the sea safe for whales. Here, the whales cannot be disturbed.
ANTARCTICA
SAFE PLACE FOR WHALES
KEEP OUT!
WHALE WATCHING
Now people go on whale-watching tours. They travel all over the world to see the whales swimming happily along.
To learn more about saving whales, send an SAE to The Whale & Dolphin Conservation Society, Alexander House, James St West, Bath BA1 2BT or visit their website at www.wdcs.org

Fast facts

Here, you can find out amazing facts about some of the whales you have met already.

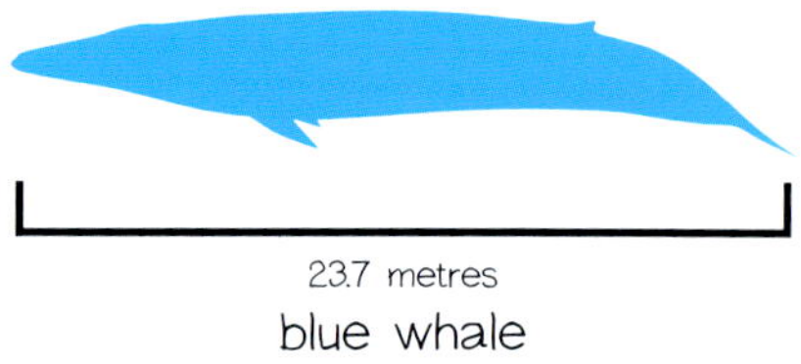

Blue whale

Weight: 109 tonnes
Teeth or baleen: baleen
Food: krill
How to spot: huge body and blue-grey skin.
Amazing skill: a blue whale blows the highest spout. It can shoot spray 9 metres into the air. That's as high as a house!
Herd size: usually swims around with two or three other blue whales.

Sperm whale

Weight: 36 tonnes
Teeth or baleen: about 50 sharp teeth, each up to 20 centimetres long.
Food: giant squid, lobsters and even small sharks.
How to spot: wrinkly skin like a prune and a gigantic square head.
Amazing skill: sperm whales dive really deep into the ocean to hunt for giant squid.
Herd size: may swim alone or in groups of 50 or more sperm whales.

Humpback whale

Weight: 27 tonnes
Teeth or baleen: baleen
Food: krill and small fish
How to spot: a pointed, bumpy snout and especially long flippers.
Amazing skill: this whale is fun to watch. It often slaps its flippers against the surface of the ocean and leaps out of the water.
Herd size: spends most of its time with one or two other humpback whales.

Grey whale

Weight: 27 tonnes
Teeth or baleen: baleen
Food: shrimp, shellfish and worms
How to spot: blotchy, grey skin covered in barnacles.
Amazing skill: grey whales swim further than any other whales. A herd of grey whales may travel up to 19,300 kilometres.
Herd size: swims alone or with several other grey whales.

Killer whale

Weight: 7 tonnes
Teeth or baleen: about 50 curved, sharp teeth.
Food: fish, birds, seals and even other whales.
How to spot: black body with white patches and large flippers shaped like paddles.
Amazing skill: killer whales are expert swimmers. They can leap out of the water and skim across the surface.
Herd size: spends its whole life in a close family group of up to 25 killer whales.

Beluga whale

Weight: 1 tonne
Teeth or baleen: about 35 small, peglike teeth.
Food: fish and squid
How to spot: white body that helps it to hide in the icy waters where it lives.
Amazing skill: beluga whales chat noisily to each other. They cheep, whistle and even roar.
Herd size: lives with between 5 and 20 other beluga whales.

Puzzles

Here are some puzzles to try. Look back in the book to help you find the answers.

Close-up!

We've zoomed in on these whales' bodies. Can you tell which parts you are looking at?

1

2

3

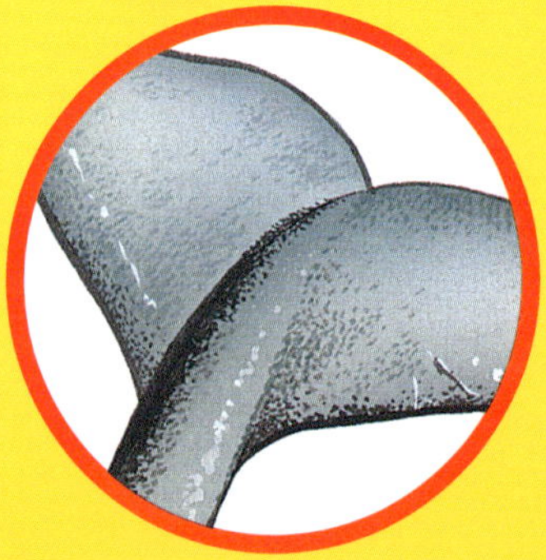

Match the babies

All of these baby whales are swimming with their mothers, except for one. Can you spot the lost baby whale swimming on its own?

Spot the difference

Look carefully at these two pictures of a bowhead whale. Can you spot four differences between them?

a

b

Who am I?

Do you know the names of these whales? Unscramble the letters to find out.

1 headbow whlea

2 hawlnar

3 backphum waleh

Answers
Close-up! 1. blowholes page 7, 2. baleen mouth page 3, 3. tail page 17 **Match the babies** The black-and-white baby killer whale showing its teeth is swimming on its own **Spot the difference** Picture b: the spout is smaller, some barnacles are missing, the whale's baleen is missing, the whale has both its flippers **Who am I?** 1. bowhead whale, 2. narwhal, 3. humpback whale

Index

Published by
Two-Can Publishing
A division of Zenith Entertainment plc
43-45 Dorset Street
London W1H 4AB

TWO CAN ™

Created by
Two-Can Design Ltd
346 Old Street
London EC1V 9RB

Main illustrations: Miles Changeur
Cartoon illustrations: Alan Rowe
Consultants: Dr Frances Dipper
Photographs: front cover NHPA/Gerard Lacz; p4 Planet Earth Pictures; p8 Bruce Coleman; p12 Ardea London Ltd.

ISBN 1-85434-803-5

Dewey Decimal Classification 599.5

Paperback 10 9 8 7 6 5 4 3 2 1

A catalogue record for this book is available from the British Library

Printed in Hong Kong by Wing King Tong